School Publishing

Do Elephants Talk?

PIONEER EDITION

By Barbara Keeler and Peter Winkler

CONTENTS

Do Elephants Talk?

The day is hot. Elephants walk across the dry ground. They are very thirsty. The **herd**, or group, has been walking for miles. They are headed to a water hole. Suddenly the elephants stop. All is quiet. But the elephants raise their ears. They seem to hear something. Yet humans hear nothing. The adults move closer to their babies. Then the elephants turn and walk away. What did they hear?

Warning Call. This elephant might make sounds warning herds to stay away from the lion.

Elephants make many sounds. They bark and roar. Their sounds can be loud. These messages help elephants keep in touch. But this time they didn't make any sounds that people could hear.

Second Language

At the water hole, the elephants made a very low sound to talk to each other. This low sound is called **infrasound**. This sound can travel for miles. It is part of a secret language. Elephants use it to talk when they are far apart.

Some messages help elephants stay safe. The elephants near the water hole heard a warning call. That is a message about danger. Another herd may have sent the message. So the herd turned away.

Translating with Technology

People cannot hear infrasound. It is too low for our ears. So how do we know about it? Scientists use machines to study infrasound. One machine records the elephants' sounds. Then another makes pictures of the sounds. The pictures stand for different elephant messages.

Hearing Aids

Elephants use infrasound to keep in touch from far away. They tune in with their ears. Their ears are good for catching sound. Elephants can stretch their ears out wide. This lets them hear many noises.

When an elephant listens, it also uses its trunk. The elephant smells the air. This may help the elephant figure out what it is hearing.

Active Listening. These elephants fan out their ears and lift their trunks when they listen.

Staying Close

Some elephant talk helps keep families together. Mothers, sisters, and children all live together. They form a herd. Males leave the herd when they are teenagers. They may live alone. Sometimes they form a herd with other males.

Young elephants sometimes wander off. They can get into trouble. Then they cry for help. Adult elephants hear the baby's cry. They use infrasound to answer. Their calls tell the calf help is on the way. Infrasound also helps them find the calf.

Making Families

Adult males and females live far apart. How do they find each other at mating time? They use infrasound. When a female is ready, she calls. Males hear the calls and head toward her.

Big Appetite. An adult can eat 300 pounds of leaves and grass in one day!

Safety Huddle. Adults huddle close to protect their calves from danger.

Elephants in Danger

Elephants can hear when danger is near. Yet elephants are still having trouble staying alive. Some people fear they may become **extinct**. This means to die out forever.

A million elephants once lived in Africa. Today, only about 500,000 live there. The number of wild elephants keeps getting smaller.

What is happening to the elephants? Some people kill them for their **tusks**. These long teeth are made of ivory.

Elephants are also losing their **habitats**, or homes. People build on land where elephants once lived.

Luckily, many people are working to save elephants. They are trying to stop the sale of ivory. They have also made parks where elephants are safe. This may help elephants survive.

Wordwise

extinct: no longer alive on Earth

habitat: the place where something lives

herd: a group of one type of animal that stays together

infrasound: sound so low in pitch that humans can't hear it

tusk: long, ivory tooth that sticks out from an elephant's mouth

Ice Baby

Long ago a baby mammoth wandered to the river. She was a month old and healthy. But this was the last day she would be alive. The baby fell into the mud and quickly sank.

About 40,000 years later, a reindeer herder found her. He lived in Siberia, Russia, and his name was Yuri Khudi. Khudi knew he had found something amazing.

Family Resemblance

Ice Baby was a woolly mammoth. She looked a lot like today's elephants, but she had small ears. Why? In Africa today, elephants are adapted to heat. They use their big ears to get rid of heat. Woolly mammoths were adapted to cold. They had small ears to keep heat in.

An Important Discovery

Khudi knew the mammoth's body was an important discovery. He went to tell a friend. Then they went to a museum and talked to the director. He told them to get Ice Baby's body.

Someone else knew the body was important. Someone had stolen it! But Khudi got it back—most of it, that is. Dogs had eaten part of its tail and ear!

The body was sent to a museum. To help thank Khudi, they named the mammoth after his wife, Lyuba.

Scientists had never seen a whole mammoth so well preserved. Lyuba's body had been sealed in mud. The mud helped preserve her body. So did chemicals from tiny organisms.

Cool Cousins. The woolly mammoth looks simliar to today's elephant.

Reading History

Scientists could tell that Lyuba had been healthy when she died. Her body was in good condition.

Scientists used Lyuba's tusks to tell about her life. Tusks grow in layers. A thick layer meant the mammoth had plenty to eat. A thin layer meant the mammoth had less to eat.

Scientists examined Lyuba. They studied her bones and her tusks and even what she had eaten. This helped them figure out what other mammoths ate at the time.

Lyuba's body helped scientists learn about Earth's history. It gave clues to what Earth was like 40,000 years ago. The plants in her stomach showed what plants were like then.

Scientists found out that Lyuba died because she had mud in her throat. She couldn't breathe.

Telltale Tusks. Lyuba's milk tusks told her age at the time of her death.

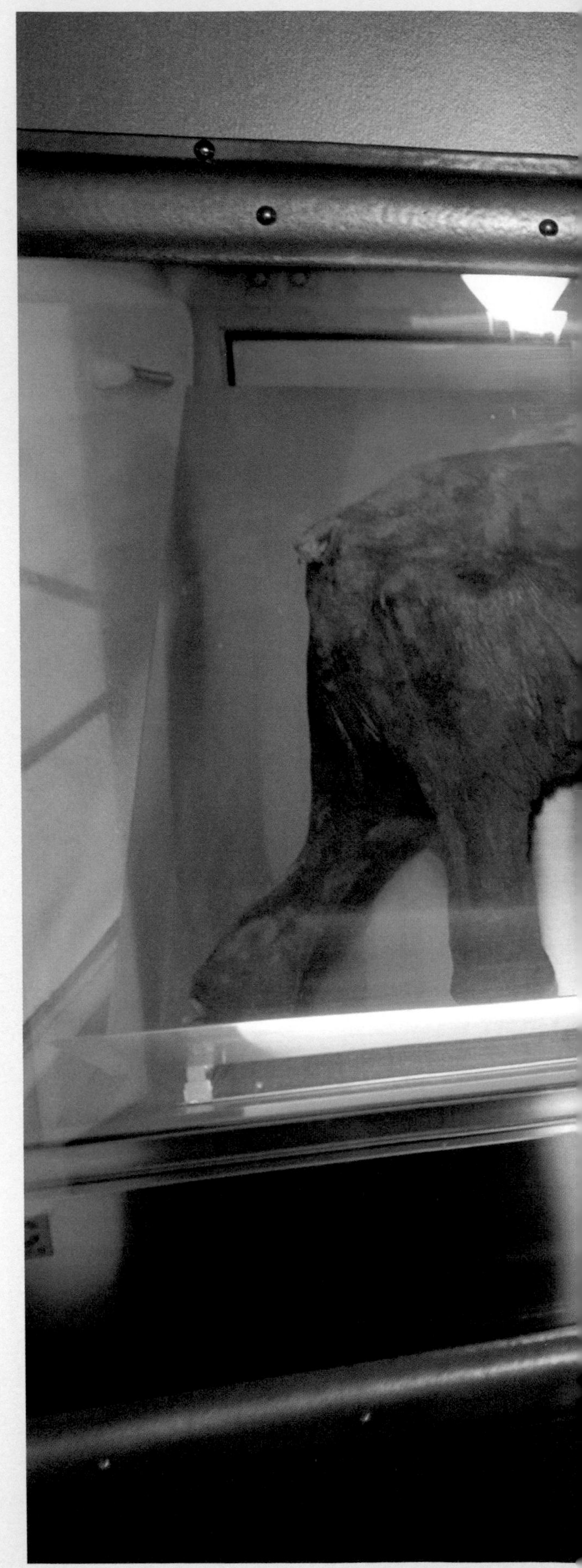

The Missing Mammoths

Most mammoths died more than 10,000 years ago. A few lived until about 4,000 years ago. Why did they die? It could be because of climate changes. Their habitats changed and they ate different plants.

Like today's elephants, mammoths lost much of their habitat. Like elephants, they were hunted by people. But mammoths may have been healthy even after the climate changed. Some scientists believe people killed off the last mammoths.

If people had tried to save mammoths, would they be extinct today? Who knows? But many people are working to protect their cousins, the elephants of today.

Perfectly Preserved. As you can see, Lyuba's body was very well preserved in the mud.

Elephants

Answer these questions to find out what you learned about elephants.

1. Why do elephants communicate with each other?
2. What parts of an elephant help it survive in its environment?
3. How are mammoths and elephants different?
4. How are elephants threatened by people?
5. What problems do today's elephants face?